Deserts & Droughts
How Does Land Ever Get Water?

Climate Basics *series*

Climate Basics series

Climate Basics: Nothing to Fear, by Rod Martin, Jr.—an Amazon #1 Bestseller in Weather and Science & Math Short Reads
Deserts & Droughts: How Does Land Ever Get Water? by Rod Martin, Jr.

Shining a Light Series

Dirt Ordinary: Shining a Light on Conspiracies, by Rod Martin, Jr.
Favorable Incompetence: Shining a Light on 9/11, by Rod Martin, Jr.
Thermophobia: Shining a Light on Global Warming, by Rod Martin, Jr.

Deserts & Droughts
How Does Land Ever Get Water?

Rod Martin, Jr.

Climate Basics *series*

Tharsis Highlands Publishing
Cebu, Philippines

Published by Tharsis Highlands Publishing
Cebu, Philippines
https://tharsishighlands.wordpress.com/books/

Amazon Print Edition
May 2019
ISBN: 9781070572741

EBook Editions
Amazon Kindle—2018
Smashwords—2018

Cover photo: Desert sunset by S.saban (CC BY-SA 4.0).
Cover design: Rod Martin, Jr.

Typography fonts
Headings: Rockwell Extra Bold
Running Heads: Rockwell
Text: Palatino Linotype

"You should not see the desert simply as some faraway place of little rain. There are many forms of thirst." — *William Langewiesche,* Sahara Unveiled: A Journey Across the Desert, *1996, Pantheon.*

Table of Contents

Introduction: Some Like it Hot

This book is all about how land ever gets water and how the climate change confusion on this topic is leading people to think some pretty crazy ideas about our climate system.

How does water ever get onto land? The simple answer is: rain. But in order to understand why rain happens, we need to look at the water cycle. That's coming up a little later in the book. But first, let's take a brief look at one of the key ingredients in the water cycle—heat.

I grew up in West Texas, which is hot enough during the summers. I did not want to have more heat. Living in Los Angeles for twenty-five years gave me plenty of mild climate. So, why would I move to Phoenix, Arizona in 1997? The answer is pretty straightforward: A job opportunity while California had become increasingly hostile to the middle class.

But Phoenix's desert climate is hot—sizzling hot. The mean maximum temperature during the summer months is about 45°C (113°F). Throughout the city, there is very little in the way of covered parking, so a car, with all the windows rolled up, becomes a blast furnace. Protection for seats and steering wheels becomes vital. After my first summer, though,

I had become acclimated to that heat. And that surprised me; I was actually starting to like it.

One thing I had missed in my decades living in Los Angeles were thunder storms. I missed the energy, the wind, the smell. I loved the lightning and thunder. With my first thunderstorm in Phoenix, I had found my arms covered with goosebumps at the first crack of thunder and then my soul felt quenched of a thirst I hadn't even known was there. The sound poured over me like life-giving rain on parched soil.

Ten years later, when I moved to the Philippines, I had become a fan of Al Gore and his film, *An Inconvenient Truth*. It spoke to my own belief in taking responsibility and our need to take better care of our world. Little did I know, at the time, that I would have a falling out with my new friend, Al Gore. I would discover that Al and I had been horribly wrong about climate.

After living in Phoenix for a decade, moving to the tropics, just 10° north of the equator, seemed downright comfortable. The humid heat was a bit different, but the temperature rarely got above 32°C (90°F). Some days, in fact, it was cool enough for long shirt sleeves. My wife thought such temperatures were cold. And sometimes, in the interior hills, it has gotten close to freezing at night.

The tropics get far more rain than Los Angeles or Phoenix. I've grown accustomed to all the life that thrives in this wetter environment. But both L.A. and Phoenix get their own share of wetness.

So, there are two kinds of heat—one dry and one wet. And on a planet with so much water, the wet kind of heat is what should interest us most, because it promotes life.

In this short book, we will look at the claims of the Warming Alarmists regarding deserts, droughts and rain, and the counter claims by those who disagree. That covers the first two chapters.

In chapter three, we look at the mechanics of how land ever gets water and why we sometimes get droughts.

And in the final chapter, we look at the empirical evidence that shows our counter claim to be true.

Note: This book contains illustrations that are black and white. For color versions of these illustrations, see the "Link to Illustrations" section of the Appendix.

For additional information on climate, see the Global Warmth blog at,

https://globalwarmthblog.wordpress.com/blog/

Chapter 1: Bogus Claims About Deserts and Droughts

Warming alarmists are fond of telling us that global warming will lead to more frequent and longer lasting droughts. Some go so far as to say that deserts are increasing in size because of manmade global warming.

Listen to the mainstream media these days and you might well be overwhelmed with the notion that the Earth is drying out like an old prune—more droughts, heat waves, brush and forest fires and burgeoning deserts. Ironically, it seems the opposite is happening, too. We hear news of too much rain with massive flooding. We also near of record cold spells, extreme winters and summer snows.

Yet, all the alarm seems to be merely wisps of smoke—here today, gone tomorrow. California, for instance, has always had droughts and likely always will. But the news-worthy drought of a few years ago suddenly disappeared and now reservoirs are above nominal levels. The extra moisture has given California a boost in plant growth. Ironically, Governor Jerry "Moonbeam" Brown vetoed a bill which would have given agencies the ability to cut back on the increased growth, greatly reducing the threat of brush and forest fires. What was he thinking? *What was he smoking?*

Recently, we had one of the largest forest fires in California history. If Brown had wanted thin justification for blaming "climate deniers," he got it by his own incompetence. Or was it incompetence? On that dark point, we simply don't know. It could just as easily have been premeditated endangerment of property and lives for political reasons. Whatever his motivation, by his actions, Brown has directly affected the destruction of property and the deaths of some of his own citizens. Some criminals go to jail for such activities, even when there was no intent (only incompetence). Some politicians seem to think they are above the law.

Part of the problem with "climate change" and the drought debate involves the term "climate change" itself. The way it is used in the media, it has built-in confirmation bias (see Appendix, Notes). Even government scientists are perpetuating this fallacy.

Another part of the problem involves the terms "scientific consensus" (including at NASA) and "settled science." Both of these act to shut down any discussion. And both of these are oxymorons (self-contradictory phrases). (See Appendix, Notes.)

And still another problem in this debate involves the very organization created to solve the so-called global warming problem. Unscientifically, they started with narrow terms of reference which define their area of concern as only manmade climate change. This is bias right up front. (See Appendix, Notes.)

News reports invariably seem to blame every drought and every flood on manmade "climate change." This implies that climate change is something new—created by industrial pollution. Contrary to what the Warming Alarmists are saying, climate has been changing for nearly 4.5 billion years—ever since Earth gained an atmosphere. Climate has been either warming or cooling and, if we believe the hysterical corporate news media, climate change has been bad for that entire period, because they never make any distinction between one kind of climate change and another. Calling the supposed problem "climate change" is thus a severe misnomer and dishonest in the extreme.

Warming alarmists repeatedly imply that modern climate change is somehow different—faster and more extreme—but it's not. There have been many times in the past when climate changed dozens of times faster and life still thrived. (see Appendix, Notes)

The movie industry has dished out one dystopic motion picture after another with themes of humans destroying the environment, frequently resulting in a desiccated landscape where water is scarce. *Blade Runner 2049* (2017), for instance, showed the interior of the United States as a deserted wasteland. This dry wasteland was also the omnipresent backdrop in *A Boy and His Dog* (1969), *Book of Eli* (2010), the *Divergent* series (2014, 2015, 2016), *Elysium* (2013), the *Mad Max* series (1979, 1981, 1985, 2015), *The Maze Runner* and sequels (2014, 2015, 2018), and *Young Ones* (2014).

Always Bad

Is "climate change" ever good? The corporate media avoids this idea.

Ann Curry of NBC reported, "Our planet is changing. Vanishing waters in the West.... There is virtually no debate among climate scientists, now. Most agree that climate change is here and that we are the biggest reason." But she's lying. Vanishing waters in the West? Virtually no debate? Most agree? Wrong, wrong and wrong again. California had a drought that lasted many months, and then the drought disappeared — replaced with plenty of rain, with winters snows in the mountains and with fuller-than-normal reservoirs.

CNN reported, "We've had more record floods, more significant drought — all these events which are more frequent and intense as a result of climate change."

The problem with their use of the "climate change" term is that they imply you would get these problems with any change in climate — warming or cooling. So, if we have bad climate right now, it becomes impossible to return to good climate, because that would require climate to _**change**_. And, as we know from all these reports, "climate change" is always bad. With this insanity, we can more easily see how the use of "climate change" as a term has been incredibly illogical.

One CBC News report stated that we need to take drastic measures to cut emissions. "If we don't, scientists say, extreme heat, droughts and floods will put hundreds of millions of lives at risk." But which scientists? Who said this? Too often, the term "scientists" refers to a nameless and faceless mob from whom we're supposed to take their words as sacrosanct. That's asking too much. Science is never done this way — or shouldn't. Transparency has always been a hallmark of good science. Scientists write papers, naming names, referencing sources and revealing their data so that other scientists can pick it apart, try it out themselves and attempt to replicate the results.

The same CNN report, mentioned earlier, painted an equally ominous picture. "Now, to a dire warning about climate change. According to a new report, experts say we have until 2030 to avoid catastrophe. It also says, if unprecedented changes are not made and made soon, there will be irreversible damage to the planet. The report focuses on what could happen if global temperatures rise by more than 1.5 degrees Celsius, or 2.7 degrees Fahrenheit. It would likely mean more erratic weather, dangerous heat waves, rising sea level and dying coral reefs.... We've had more record floods, more significant drought—all these events which are more frequent and intense as a result of climate change."

This CNN report is full of journalistic problems. Most apparent is its appeal to emotion (a common logical fallacy) with words like "dire," "avoid catastrophe," "unprecedented," "irreversible damage," "erratic," "dangerous," "dying," and "more frequent and intense." They use the ubiquitous and nameless "experts" which is an appeal to authority (another logical fallacy). Their mention of "irreversible damage" is wrong. The Warming Alarmists blamed corals dying out on warming, but the largest batch of corals—the Great Barrier Reef—is recovering quite nicely; corals have existed long, long before our current Ice Age began, thriving in far warmer temperatures. Their use of the word "likely" is also problematic. Likely by who's estimation? How likely? And what criteria were used to determine the degree of likelihood? Nothing they said answers these critical questions. We're supposed to take their word for it all. The claim "more frequent and intense" is demonstrably false. More on that later in the book.

ABC Australia adds to this dark picture of our future. "Human activities, from pollution to overpopulation, are driving up the Earth's temperature, and fundamentally changing the world around us.... The main causes are the phenomenon known as the greenhouse effect.... The rapid increase of greenhouse gases in the atmosphere, has warmed the planet at an alarming rate.... Warmer temperatures also make weather more extreme. This means not only more intense major storms, floods and heavy snow fall, but also longer and more frequent droughts.... Growing crops becomes more difficult.... and water supplies are diminished."

Like CNN, ABC Australia is being sloppy with their language. Warming from pollution? Here it seems they're referring to CO2—a greenhouse gas. But carbon dioxide is not, and never has been, a pollutant. It remains a vital gas of life without which all life would die. And as far as having too much CO2, we're still on the side of carbon dioxide starvation from too little. The correlation between CO2 and temperature is accidental; not cause-and-effect, except where temperature changes force carbon dioxide into and out of the oceans.

Overpopulation is far from a problem, despite what centuries of doomsayers have been talking about (see Appendix, Notes).

Ian Johnston wrote in the Independent, "Professor Michael Mann said extreme weather events—such as the 'unprecedented' drought in California last year, the flooding in Pakistan in 2010 and the heatwave in Europe in 2003—were happening more often than they should do, even taking the warming climate into account." The problem with Michael "Hockeystick" Mann's hyperbolic comments is that they

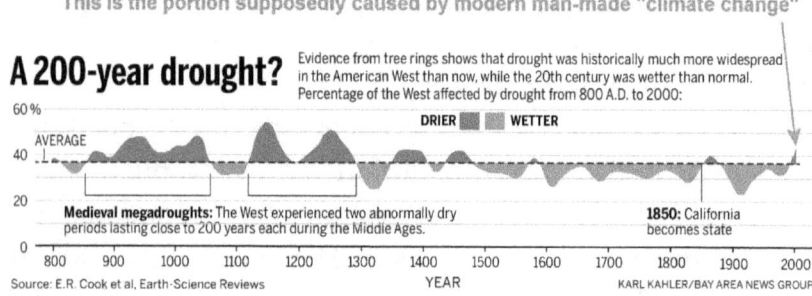

Figure 1.1—This graph was developed from current and proxy drought data while the California drought was still ongoing. Notice the size and duration of the previous droughts.

remain half-truths and outright lies. The recent years of drought in California were not "unprecedented." There have been far longer and far deeper droughts in the paleoclimate record.

Are droughts and heatwaves really happening more often? One look at the above graph and you know how wrong Mr. Mann is on that point. Heatwaves? In 1850, our world came out of a 500-year cold spell called the Little Ice Age; of course it's going to be warmer! But in the context of all history, the Modern Warm Period of the 20th and 21st century is a tiny bump compared to the mountains of warming which came before. In fact, during merely the Holocene interglacial of the current Ice Age, the Modern is the coldest of the Holocene's 10 major warm periods (1,000-year cycle).

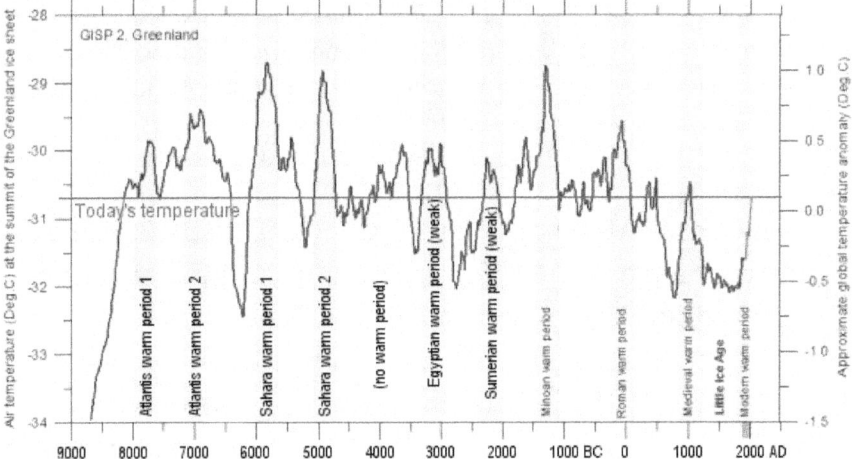

Figure 1.2—A graph of the Holocene temperature from GISP2 (Greenland) ice cores. Notice the scientist's label on the right—"Approximate global temperature anomaly (Deg. C)." The original graph has been modified by adding the recent temperature, in red, following the ice core proxy (blue), "Today's temperature" in red, and an additional 7 bars of green after the 3 most recent, to show the periodicity of the warm periods.

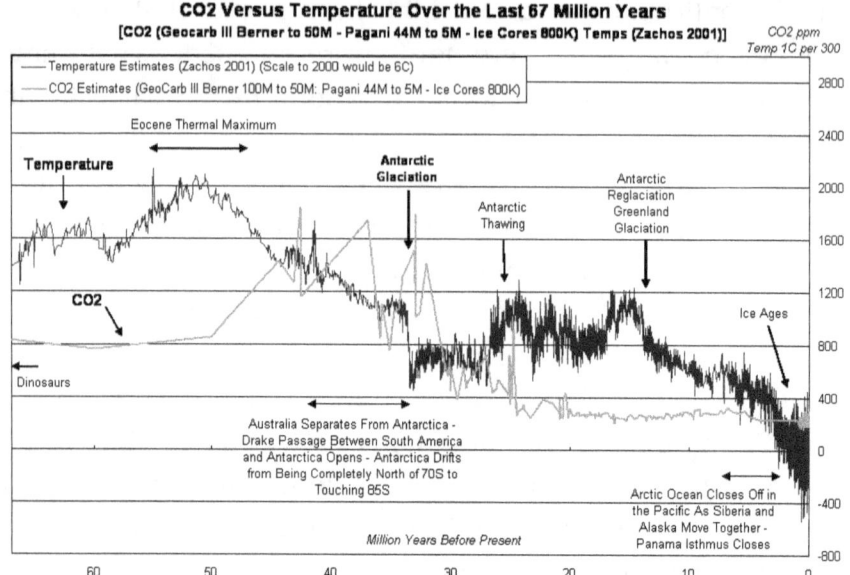

CO2 Versus Temperature Over the Last 67 Million Years
[CO2 (Geocarb III Berner to 50M - Pagani 44M to 5M - Ice Cores 800K) Temps (Zachos 2001)]

Figure 1.3—Earth's climate over the last 67 million years with the 2.6-million-year Pleistocene Ice Age on the far right. The prior 3-million-year Pliocene, and 17.7-million-year Miocene Epochs had warm periods that were far warmer than at any time during the current Ice Age. In fact, only a couple of brief times did the Miocene "cold" dip down to the level of the Holocene's warmest periods.

Michael Mann and James Gleick in their 2015 paper, "Climate change and California drought in the 21st century," wrote, "California is experiencing extreme drought. Measured both by precipitation and by runoff in the Sacramento and San Joaquin river basins, 10 of the past 14 y have been below normal, and the past 3 y have been the driest and hottest in the full instrumental record." Notice the last few words, "in the full instrumental record." This means "ever since man-made instruments were created for measuring such things." But that claim avoids the fact that we have moderately reliable proxy data that goes back far further than the age of instrumentation. It also avoids the fact that we had just come of the Little Ice Age—500 years of colder climate. The proxy data shows that droughts have been far worse in California's past, as shown by the graph, above (Fig. 1.1). But their statement also avoids the glaring reality that any measurement on the warming side of any major warming period peak is going to be warmer than most data since the coldest part of the previous cold period valley. It's like they're saying on June 22nd of any year, this day is far hotter than any day this year. Coming out of winter tends to do that.

The level of childish simplicity and ignoring context is outrageous from the scientists, politicians and corporate media pundits. Don't get me wrong; I appreciate simplicity, but only when it is in the full context of the larger picture. Simplicity that trims out important facts is misleading at best.

For PBS, Michael Mann stated, "What we can conclude with a great deal of confidence now is that climate change is making these events more extreme. And it's not rocket science. You warm up the atmosphere, it is going to hold more moisture, you get larger flooding events, you get more rainfall.

"You warm the planet, you're going to get more frequent and intense heat waves. You warm the soils, you dry them out, you get worst drought. You bring all that together, and those are all the ingredients for unprecedented wildfires."

Notice how Mann uses the term "climate change." This "Swiss-army-knife" term is used for a host of ills—especially both droughts and floods—two contradictory phenomena—dryness and wetness. Climate change also includes cooling, yet Mann misuses the term to include only warming. Notice also how he doesn't say "worse drought," but "worst drought," and this is demonstrably wrong.

The Union of Concerned Scientists had this to say on drought, "Global climate change affects a variety of factors associated with drought. There is high confidence that increased temperatures will lead to more precipitation falling as rain rather than snow, earlier snow melt, and increased evaporation and transpiration. Thus the risk of hydrological and agricultural drought increases as temperatures rise." They warn that we'll get more rain instead of snow and that this is somehow bad because of increased evaporation. We'll soon see how their description is horribly wrong, because it doesn't matter how quickly water evaporates if you get more rain to replace the amount evaporated. It doesn't matter that the surface of land is dry for part of the day if it is quenched of its thirst every day.

Jeremy Deaton, writing for Popular Science, concerning California's lack of abundant snowfall, stated, "Blame it on climate change...

"The root cause of all this mayhem is climate change. Carbon pollution is trapping heat, which is cooking the planet. Warmer water temperatures in the Pacific Ocean are giving rise to atmospheric rivers that deliver rain — instead of snow — to the Sierra Nevada. Heavy rainfall threatens to melt what little snow gathers on the slopes. This has consequences for the entire state, as reduced snowpack fuels drought more broadly, yielding wildfires and mudslides in coastal areas as well as in the mountains.

"In a recent op-ed for the Los Angeles Times, UCLA climate scientist Alex Hall and science communicator Katharine Davis Reich warned, 'In simple terms: We're going to lose a lot of snow to climate change. Equally worrisome, California's water infrastructure is not resilient enough to make up for the loss.'"

Again, the specter of "climate change" rears its ugly head. They don't blame it on global warming, but on this non-specific boogie man, "climate change."

Another part of the problem involves the conflation of climate models with reality. They're not the same thing. Yet, organizations like Science Daily perpetuate this myth with headlines like, "Climate change intensifies droughts in Europe." Their subheading exacerbates this problem: "Researchers model the effects of the global temperature rise." There is little to differentiate between reality and models. The title acts as though this is the reality; the subheading makes it sound like the models are perfectly accurate. They aren't. Their article goes on to say, "According to the modelling results of the author team -- which involved scientists from the USA, the Netherlands and the United Kingdom in addition to the UFZ -- if global warming rises by three degrees, the drought regions in Europe will expand from 13 percent to 26 percent of the total area compared to the reference period of 1971 to 2000." Please notice the word "if." That pops their delicate balloon of pretense, and it deflates the impact of their title.

The overall problem with this claim of droughts and growing deserts is that it remains horribly wrong. It's a lie. Sure, there will be changes. Some areas will become drier; some wetter. That has always happened. Things like this are to be expected in a non-linear, chaotic system like global climate. But overall, greater warmth means fewer droughts and smaller, tighter deserts.

Chapter 2: Counter-Claim — Deserts and Droughts Diminish from More Warming

Global warming results in fewer, shorter droughts and smaller deserts. Saying it, alone, doesn't make it so. But keep reading. You will find a host of facts that prove the validity of this counter-claim. In this chapter, we look at what others say on the topic. In the next chapter, we look at the mechanics of the water cycle, and in the final chapter, we dig into the empirical evidence.

The entire global warming hysteria is not based upon science, but emotion. Humanity suffers the famines of the Little Ice Age and then is blessed with another major warm period on the 1,000-year cycle of our Holocene interglacial. Quite a number of scientists have a great deal to say about the unscientific nature of the "climate change" movement. Among them is MIT climatologist Dr. Richard Lindzen. ReasonTV recorded a lecture Dr. Lindzen gave on the topic of The Politics of Global Warming. In that lecture, Lindzen stated, "Being skeptical about global warming does not, by itself, make one a good scientist. Nor does endorsing global warming make one, *per se,* a poor scientist.

"And one of the most difficult things, I think, for someone who is actively involved in the scientific community is to realize, in my case for instance, that most of the atmospheric scientists who I respect, do endorse global warming. The important point, however, is that the science that they do that I respect is not about global warming. Endorsing global warming just makes their lives easier."

It remains understandable that even scientists can be selfish enough to accept garbage science while producing good science. This happened in Nazi Germany, where scientists who disagreed too vocally with the bad science of government propaganda ended up arrested or dead. Today's insanity hasn't yet gotten that bad, but talk in the news and social media has some crying out for "deniers" to be arrested or killed. Bill Nye, the funny bow-tie guy, was quite willing to have the opposition jailed for their contrary views.

Lindzen went on to say, "The process of co-opting science on behalf of a political movement has had an extraordinarily corrupting influence on science, especially since the issue has been a major motivation for funding. Most funding for climate would not be there without this issue.... The global warming issue has done much to set back climate science. In particular, the notion that climate is one-dimensional—which is to say that it is totally described by some fictitious global mean temperature and some single, gross forcing—*à la* increased CO_2—is grotesque in its oversimplification."

On the topic of Little Ice Age Europe, C.E.P. Brooks wrote in his book, *Climate Through the Ages,* "The period 1701 to 1750 comes out as unusually dry." The reason is easy to understand. Colder oceans evaporate less water so rain will naturally become more scarce. This doesn't mean all areas experience less rain. It only means that the planet as a whole will receive less rain, because there will be less water vapor available to become rain.

Definition from Random House Encyclopedia:

"Drought, a condition that occurs when evaporation and transpiration exceed precipitation for considerable periods. Four kinds are recognized: permanent, typical of arid and semi-arid regions; seasonal, in climates with a well-defined dry and rainy season; unpredictable, an abnormal failure of expected rainfall; and invisible, when even frequent showers do not restore sufficient moisture lost to evaporation."

Helmut Landsberg adds to our understanding of the topic in his textbook, *Physical Climatology*. Landsberg was, at the time, Director, Office of Climatology, United States Weather Bureau.

"Drought. Strictly speaking drought is not a physical but a biological phenomenon. It reflects complicated relations between the plant, soil, and atmospheric conditions. Each plant type has its own responses to moisture deficit. Some are extraordinarily drought-resistant, others wilt quickly. In nature, undisturbed by human activity, species of plants are indigenous to an area or migrate there when the climate in all its ramifications and variations is suitable for survival. Thus each major climatic zone has its own characteristic plant associations. Even the arid regions have some vegetation. Certain grasses, cacti, and sagebrush can get along on a minimum of water. Aridity is a natural stage but drought is primarily an affliction of cultivated regions. It results often when crop or ornamental plants are introduced into a region to which they are not completely climatically adapted. While they may get along in the best years, they perish in the worst seasons which the natural vegetation manages to survive."

Notice how Landsberg describes climate change in this short dissertation on drought—how plants "migrate... when the climate in all its ramifications and variations is suitable for survival." So, even in the 1950s, climatologists were well aware that climate changes, making a region more or less hospitable to a type of plant.

Gabriel Henderson, in his 2014 article, "The Dilemma of Reticence:Helmut Landsberg, Stephen Schneider, and public communication of climate risk, 1971-1976," quoted Landsberg, "Science is not as objective as some people think. Often human value judgments (or even prejudices) make things move as much as curiosity or the search for answers as to 'why.'"

Henderson went on to characterize Landsberg's viewpoint on the new controversy of "climate change." "Landsberg cast considerable doubt on the validity of relying on computer-based models to inform policy makers and the general public. Unless one could adequately quantify the scientific uncertainties that underlay scientific claims based on models, he believed that reticence was the only appropriate course of action until such uncertainties could be identified and resolved. Staying behind closed doors, cautiously hedging one's claims by quantifying and emphasizing scientific uncertainty, and diligently collecting and analyzing data to resolve such uncertainties were hallmark characteristics of what he envisioned to be a professional atmospheric scientist."

Specific Cases of Drought—California and Texas

People were freaking out in California during its recent, multiyear drought. But we need to learn to take such things in the proper context. Jeannette Warnert, writing in 2014 for the University of California Division of Agriculture and Natural

Resources, said, "Scientists studying long-ago California climate have realized that the 20th century was abnormally wet and rainy, according to researcher Lynn Ingram, professor in the Department of Earth and Planetary science at UC Berkeley. 'The past 150 years have been wetter than the past 2,000 years,' Ingram said. 'And this is when our water development, population growth and agricultural industry were established.'...

"Precipitation during the last three years in California has been low by standards set since records were kept, which began in the late 1800s. However, the current drought appears to be well within normal fluctuations in the state's climate, according to research by Ingram and other paleoclimatologists."

Warnert went on to describe the work of Scott Stine of California State University East Bay. She said that he "...found some of the first evidence of a medieval warm period in California by studying the water level of Mono Lake. The lake expands and contracts depending on the amount of runoff from the adjoining Sierra Nevada. Stine's research reveals a dry spell from 1,800 to 600 years ago" or 200–1400 AD. This shows that some areas get drier during a warm period. But don't be fooled by this. Any change in climate will cause problems somewhere. And climate always changes. So, we need to be alert to the changes and to take proactive steps to prepare in advance for the problems.

Regarding drought conditions in Texas, as of November 15, 2018, less than 2% of the state was experiencing minor levels of dryness or a moderate level of drought, while 98.46% of the state was experiencing no drought conditions at all.

The National Drought Mitigation Center provides us with clear pictures of the drought situation in the United States, including time series charts and drought maps. Here's the drought timeline for Texas over the last 18+ years:

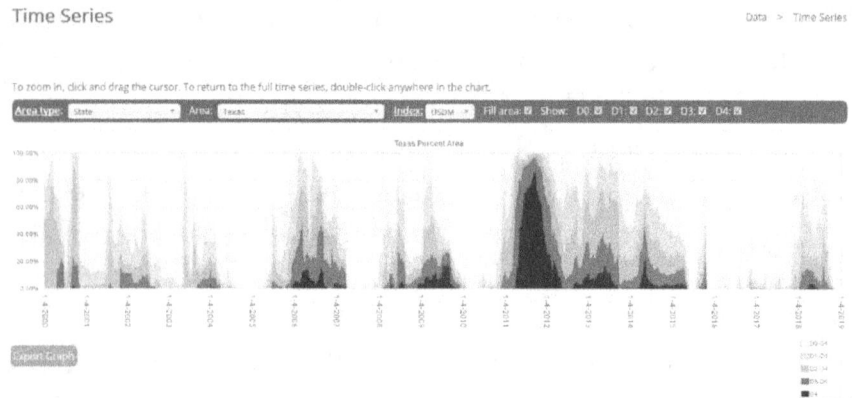

Figure 2.1—Here is the timeline of drought conditions for Texas from the year 2000 to 2018.

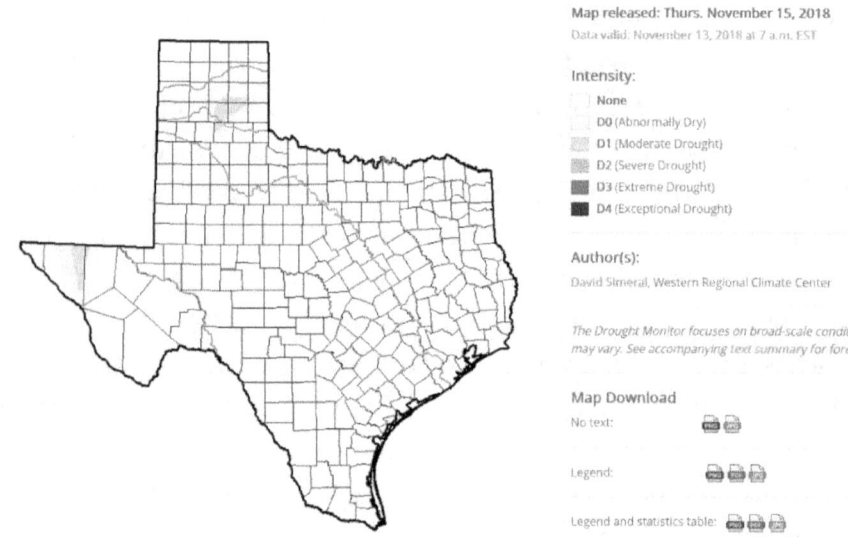

Figure 2.2—This is a map of Texas showing the areas suffering a tiny amount of dryness and drought.

https://droughtmonitor.unl.edu/CurrentMap/StateDroughtMonitor.aspx?TX

As you can see from the timeline (Fig. 2.1), Texas had a severe drought for most of 2011 and the beginning of 2012. But like virtually all droughts in areas which normally have few, if any droughts, the period of extreme dryness was short-lived.

All of the hysteria, gloom and doom was wasted effort and worthless bluster.

The Grand Cooling of Earth

The Eocene Epoch (56.0–33.9 Myrs ago) was one of the warmest periods in Earth's history. After Australia separated from Antarctica, ocean current patterns changed significantly, leading to a prolonged period of cooling. With that cooling, Earth dried out. This is common knowledge throughout the geological and paleoclimate literature.

The University of California Museum of Paleontology says this about the "grand cooling" period: "By the Late Eocene, the new ocean circulation resulted in a significantly lower mean annual temperature, with greater variability and seasonality worldwide. The lower temperatures and increased seasonality drove increased body size of mammals, and caused a shift towards increasingly open savanna-like vegetation, with a corresponding reduction in forests."

The increasing dryness led to the elimination of some forests, replacing them with grass and brush.

Even Wikipedia, which frequently leans toward the Warming Alarmist position, backs up this idea in its discussion of the Eocene Epoch: "At the beginning of the Eocene, the high temperatures and warm oceans created a moist, balmy environment, with forests spreading throughout the Earth from pole to pole. Apart from the driest deserts, Earth must have been entirely covered in forests."

Scientists have long known that cooling climate leads to a drier environment. In *The Planet We Live On: An Illustrated Encyclopedia of the Earth Sciences* (ed. C.S. Hurlbut), we learn about this cooling in the article on the Oligocene Epoch (33.9–23.03 Myrs ago; the period right after the Eocene Thermal Maximum): "Oligocene climates were somewhat cooler and drier than the preceding Eocene Epoch, but fossil palms in northern Germany and the remains of alligators and palms in the Dakotas suggest warmer climates for these areas. Their article on the Miocene Epoch (23.03–5.333 Myrs ago) gives us yet another example of this effect: "General uplift of the northern continents and localized mountain building produced cooler and drier climates in much of the world, and salt deposits in southern Nevada suggest desert-like conditions for that part of North America."

We find evidence even on ebay.com of the wetter nature of the warmer and earlier Eocene climate—fossilized pine cones from the Sahara during the Eocene. Pine cones in the Sahara?

From the far warmer climate which saw virtually no polar ice—at least none that persisted throughout the year—Earth cooled down, and as it did, it dried out, too. We'll talk about why this happens in the next chapter, but for now we need to look at the fact that the opposite effect would occur if Earth warmed instead of cooled—warming leads to a warmer, greener environment over most of the planet.

Chapter 3: How it All Works — Rain, Droughts and Deserts

Water is a most wonderful substance. Not only is it an essential ingredient for life, but water helps to manage the climate of the entire planet through its three phases—solid, liquid and gas.

With water, temperature is kept from getting too hot or too cold. It takes a great deal of heat energy to evaporate water, and it takes a similarly great deal of heat energy loss to freeze water. These tend to buffer the climate, keeping it from runaway warming or freezing.

Where there is plenty of water, daily temperature swings are relatively slight. Where there is very little water, daily temperatures can change quite dramatically—scorching during the day and freezing during the night.

On our key question—How does land ever get water?—we've already looked at the simplest answer: rain. But how does rain work?

It all starts with warmth from the sun. Warmth gives water the energy needed to evaporate and to become water vapor—a clear, odorless gas. The warmer the air, the more water vapor it can hold. When air becomes saturated with water vapor, it tends to form fog or clouds—tiny droplets of water that condense out of the water vapor.

When these droplets become big enough, they can no longer be suspended in air and fall toward the ground as rain or snow.

Warmth Drives the Water Cycle

Evaporation and precipitation are the two key parts of the water cycle, but there are other steps. Wind can help with evaporation by taking saturated air away from the water surface, allowing new evaporation to take place. In the sky, water vapor requires particles of dust or other methods to affect nucleation—the formation of cloud droplets. Once formed, cloud droplets can grow as cool air forces more water vapor to condense.

The other process involves the cloud chamber effect, where fast moving atomic particles, called cosmic rays, trigger cloud nucleation. With cosmic rays, dust particles are not needed.

Naturally, more dust or more cosmic rays can result in more clouds being formed. Volcanic eruptions and dust storms can seed water vapor in saturated and super-saturated air. And when the sun emits less solar wind, more cosmic rays can reach Earth to trigger cloud formation.

As hot air rises, it cools. Any parcel of air raised in altitude will cool, at least anywhere in the troposphere (the lower and thickest portion of the atmosphere). When warm, moist air is forced up a mountain slope by winds, it cools and forms clouds. This is one big reason why mountains on the coast frequently have more rain and plant life. Mountains and valleys farther inland are drier and less hospitable to life.

Evaporation cools the surface of the Earth. Clouds also cool the Earth's surface by blocking sunlight and sending it back into space by reflection. These are two key methods by which water regulates our planet's temperature.

In cold regions, the falling water comes down frozen, either as snow or hail. Over most of the planet, the water comes down as rain, wetting the ground, frequently running off into streams or lakes, or falling directly onto the ocean. Streams converge forming rivers which usually empty into the ocean and sometimes inland seas.

With Greater Warmth, We Get More Rain or Snow

When we have fewer clouds, we get more sunlight hitting the surface, which directly raises the temperature at or near the surface. Where there is water, the warmth causes some of the liquid to evaporate, turning it into water vapor. The energy required for the phase change from liquid to gas ends up cooling the surface. Between the two forces—evaporation and sunlight—an equilibrium is achieved which determines the resulting temperature.

Warm, moist air rises into the sky, forming clouds. Sometimes, the conditions are right for the formation of *cumulonimbus*—rain clouds.

More warmth generates more rain or other precipitation. The opposite is true, as well. With cold enough climate, there would be virtually no evaporation and no precipitation. This is why the polar regions are considered to be the world's two largest deserts.

Global warming will result in more rain—not less. More rain means fewer droughts and smaller deserts. That additional rain won't affect every location on Earth, but it will expand the reach of rain into regions that were previously considered desert or semi-desert. The overall effect is to upgrade the region to one supporting more life. Ironically, some Warming Alarmists consider this upgrade to be something bad for the planet, as if all change is detrimental. It isn't. Life is not the villain; more life is a good thing.

Some Warming Alarmists have even gone so far as to argue against the "more warmth—more rain" paradigm. They state that warmer air can hold more water, so there won't be more rain.

While it's true that warmer air can hold more water vapor before becoming saturated, once it becomes saturated, any additional water vapor pushes the air into a super-saturated state which can easily lead to precipitation with the presence of dust or cosmic rays.

But let's do a thought experiment to explore more thoroughly this erroneous notion of an increased evaporation *not* leading to more rain. Let us say that we currently have 2 units of evaporation every day from the oceans. This would result in 2 units of precipitation. Now, let's say that fewer clouds allow the surface to become warmer, generating 3 units of evaporation. After the first day, we would have experienced 3 units of evaporation, but only 2 units of precipitation. The net effect would be a gain of 1 unit of water vapor left in the sky. After a week at this new rate, we would have an additional 7 units of water vapor stuck in the sky. See where this is going? Eventually, we would end up with the entire ocean above our heads. This clearly illustrates how wrong their argument is. Nature always finds a balance at a new rate in the water cycle. Nature abhors imbalance. So, after saturation is reached, the amount of precipitation will match the amount of evaporation.

The Hadley Cell and Planetary Circulation

Hot air rises and cold air descends. As air rises, it cools significantly. Warm moist air, rising from the tropics, drops most of its moisture on the tropics and heads poleward. Around the Tropics of Cancer and Capricorn, 23° 26′ 12.6″ latitude, north and south respectively, the cooled tropical air descends in what are called the subtropics. As it descends, it heats up. Warmer air can hold more water vapor, so the descending air becomes increasingly dry because it is further from the saturation level for that temperature. Remember, the moist tropical air had already dumped much of its moisture in the tropics, watering all that green plant life.

This is why the lands around 23° north and south latitude are mostly desert. Here, you find the great deserts of North and South Africa—the Sahara and Kalihari. We also have the Arabian Desert in Saudi Arabia, the Thar Desert in India, the deserts of Mexico and the American Southwest, plus the great desert outback of Australia which covers a large percentage of the island continent. The air that comes down in the subtropics then moves along the surface, back toward the equator. This massive circulation is called the Hadley Cell— one for the northern hemisphere and one for the southern.

Between the subtropics and about 60° north and south latitude, we have similar cells each called the Ferrell Cell. And between 60° and the poles, we have two more, each called the Polar Cell.

Warmer air rising from around 60° latitude drops its moisture as rain or snow, with some going poleward, and some going toward the subtropics—where the drier air descends, becoming even drier as it warms.

Deserts and Droughts

Some deserts don't fit the simple, Hadley Cell placement mentioned above. The Gobi Desert in China and Mongolia stands at about 43° north latitude—the same latitude of forested Vermont, New Hampshire and Oregon.

Sometimes prevailing winds blow moist air in certain directions and not in others. Coastal mountains can suck the moisture out of air, so that the leeward side is left dry and largely barren. We see this effect in Washington and Oregon where the coastal lands are rich in forests, while the eastern side of each state is semi-desert or grassland. The deep interior of continents is frequently drier, because they are farther from the sources of moisture (oceans and seas) and rain clouds have already dropped their loads before reaching those inner lands.

The polar regions are considered deserts because they receive very little precipitation. All that accumulated snow can be misleading. There is so much of it only because most of it is thousands of years old; it never melted. The polar regions are doubly deadly because they are both cold and dry. Contrast this with the tropics which are full of life and biodiversity, because they are both warm and wet.

Droughts occur when regions which normally get some rain suddenly receive less than their usual. This can happen when weather patterns shift the warm, moist air away from those regions. Droughts always happen, because the climate system is a set of non-linear, chaotic phenomena.

Climate is always changing. In fact, climate has never ceased to change ever since Earth gained an atmosphere, nearly 4.5 billion years ago.

What we know is that with greater warmth, we have more rain, and this leads to fewer droughts and smaller deserts.

Climate cycles typically repeat until a great enough force changes the elements that control those cycles. When Australia broke off from Antarctica and they became two separate continents, warm temperate waters were no longer forced toward the south polar region. A circumpolar circulation was established which effectively cut Antarctica off from the sources of warmth. And that's when Antarctica lost its forests and started gaining its glacial covering—millions of years of light snowfall, accumulating into mountains of ice.

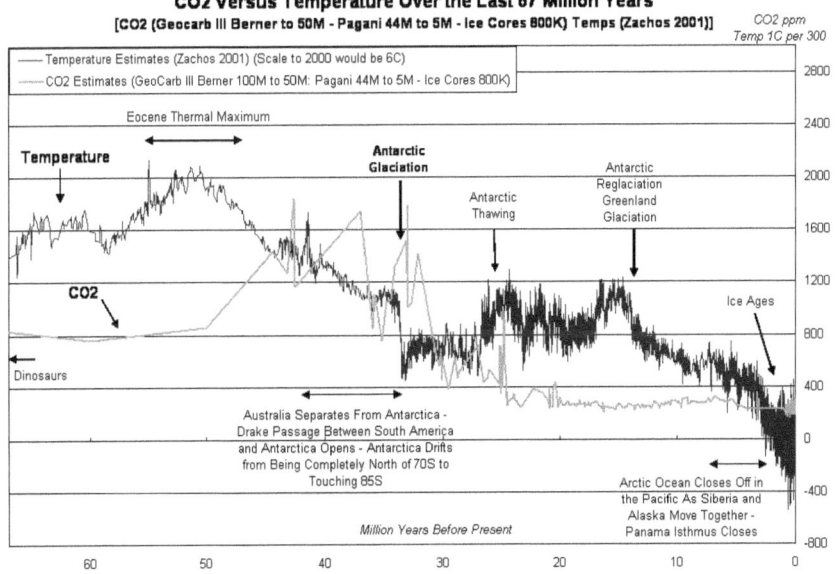

Figure 3.1—Notice the massive cooling about 34 Mya, followed by 8 million years of Antarctica glaciation. That cooling turned Antarctica into a desert.

With this theoretical understanding, it seems clear that the Warming Alarmists' claims about global warming and droughts are wrong. But what does the evidence show? We will look at this in the next chapter.

Chapter 4: Empirical Evidence About Deserts and Droughts

Recent Alarming Droughts are Puny Compared to Those of the Past

For all the corporate media hype, recent droughts were nothing compared to those in the more distant past. The alarm bells of death and destruction from "man made climate change" have been greatly overblown.

Take for instance the California drought of the early 2010s. Toby Ault and Scott St. George, in a recent article for Physics Today, talk about megadroughts of the past that make the recent California drought look like the proverbial picnic. "...the multiyear drought that first gripped California in 2012 officially ended only last year [2017]. Those events were among the most severe and long-lasting dry spells of the past 150 years. If we look back further, however, evidence from natural climate archives suggests that the recent droughts were eclipsed in previous centuries by long-lived events known as megadroughts." The article's subheading states, "During the first half of the second millennium, the American West was plagued by a succession of decades-long dry spells." The graph in chapter one on megadroughts of the American

West showed us that one such drought lasted nearly two centuries and reached far deeper levels of desiccation.

So, this notion that humans are driving the planet toward oblivion is sheer nonsense. Recent droughts are neither very big or long, and certainly not unusual.

Upside-Down Claims

Ironically, in the 1970s, scientists were blaming California drought on global cooling. Michael Bastasch wrote in 2015, "California was stuck in a deep drought during Gov. Jerry Brown's first term, much like the one the state is currently going through. The only difference is that global cooling, not warming, was blamed for causing drought in the late 1970s.

"In 1976, the New York Times reported that California was 'so dry, brush fires have started several weeks early' and that 'water is being rationed.' But in the 1970s, scientists blamed this drought on global cooling.

"The Times reported that climatologists 'believe that the climate has moved into a cooling cycle, which means highly erratic weather for decades to come.' Scientists worried that the world's population had gotten so high that minor 'shifts in climate could be catastrophic.'"

What is doubly ironic about this assessment is that some of the same descriptions used to make us afraid of global cooling are being used today to have us fear global warming—erratic weather, droughts, water rationing and large populations in danger.

The Great Imposter

When someone makes claims that two opposing phenomena will have the same end result, we have to be wary that they're trying to pull something over on us. One, or both, of those claims is dead wrong—an imposter. Consider, for example, the claims that hot air rises and cold air rises. One of them is wrong!

In a short article published earlier this year, Tony Heller dug up and displayed a news clipping from 1981 which proves his title, "SCIENCE : Symptoms Of Global Cooling And Global Warming Are Identical."

From the *Chicago Tribune*, Wednesday, November 25, 1981, "Many authorities on population growth and food supply now fear that the world is close to the brink of a Malthusian disaster. In fact, several ominous developments in recent years have brought us closer to the edge of this abyss.

"Climatologists now blame recurring droughts and floods on a global cooling trend that could trigger massive tragedies for mankind.

"Population growth in Third World nations is raging unchecked as it continues to outstrip gains in food production. It has been estimated that one-fourth of all the human beings born since the dawn of recorded history are alive today."

A snapshot of internet search results gives similar claims about global warming:

- "Man-made global warming makes droughts and floods more likely." [engadget.com]
- "Why does climate change lead to more floods and droughts?" [climaterealityproject.org]
- "Global Warming Impacts | Union of Concerned Scientists." [ucsusa.org] "An overview of the impacts of global warming, including sea level rise, more frequent and ... National Landmarks at Risk: Rising Seas, Floods, and Wildfires Are ... As temperatures have warmed, the prevalence and duration of drought has ..."
- "Climate change is altering global air currents — increasing droughts." [independent.co.uk]

- "Droughts, Floods, and Wildfire — Climate Science Special Report." [science2017.globalchange.gov]
- "Extreme Weather | National Climate Assessment." [nca2014.globalchange.gov]

They can't have it both ways. One claim is dead wrong. Part of the difficulty with this topic is that any change in climate will cause problems somewhere in the world. The climate and weather are chaotic forces that are relatively stable, globally, but locally unpredictable. You're always going to have droughts and floods, but focusing news reports on them skews perception greatly. Just because we have droughts (and also floods) doesn't mean that global warming or cooling are causing them.

In the 1970s, 100% of the national news about Los Angeles for one particular day was all about massive rains, floods, mudslides and houses falling off of hillsides. The alarming nature of the news was so intense that one of my relatives called me up long distance to make sure I was okay. The news was really only about far less than 1% of reality. Perhaps as few as 10 houses suffered catastrophic collapse, and several million homes suffered nothing. If all you see on the news are floods and droughts, then you have no per-spective—no broader context. That's why we need to measure the amount of any phenomenon and compare it over time. Without this, everything can look bad, and that's far from accurate. That's how floods and droughts can be blamed on both global cooling and global warming. All of the evidence presented in the corporate news media is anecdotal.

As things cooled off toward the current Ice Age, scientists know that the environment dried out.

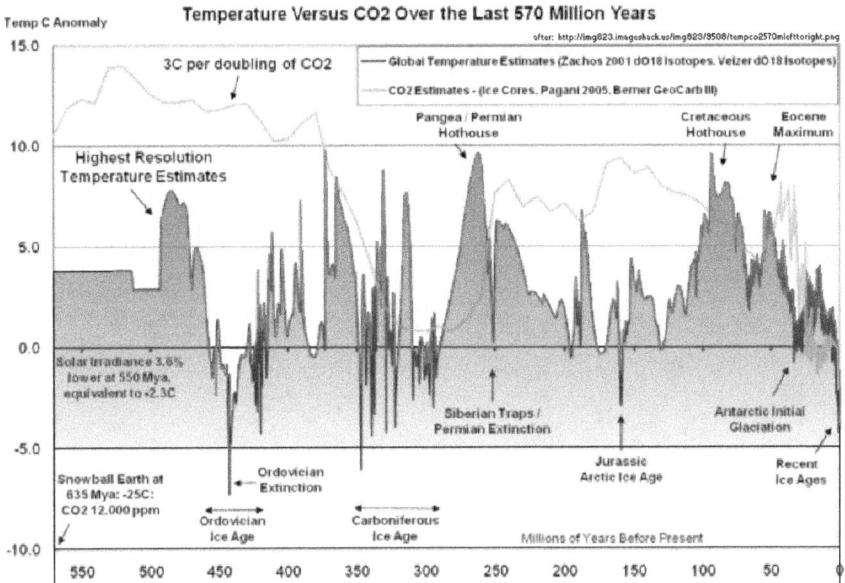

Figure 4.1—Graph of temperatures and CO2 for the last 570 million years. Red shading has been added to the original graph in order to amplify the range of temperatures involved.

Key: Cooler = Drier! Warmer = Wetter!

Cooler = drier. So, logically it follows that warmer = wetter. And it shouldn't need stating explicitly, but wetter = fewer, smaller droughts and deserts. Incredibly, some scientists are on record saying the opposite. How they can say such things with a straight face seems even more incredible. It's like they're saying "cold air rises" (wrong) and "hot air sinks" (wrong, again). Most humans, if they took a moment to think carefully about this, would realize that they were being told a lie.

National Drought Mitigation Center:

Time Series   Data > Time Series

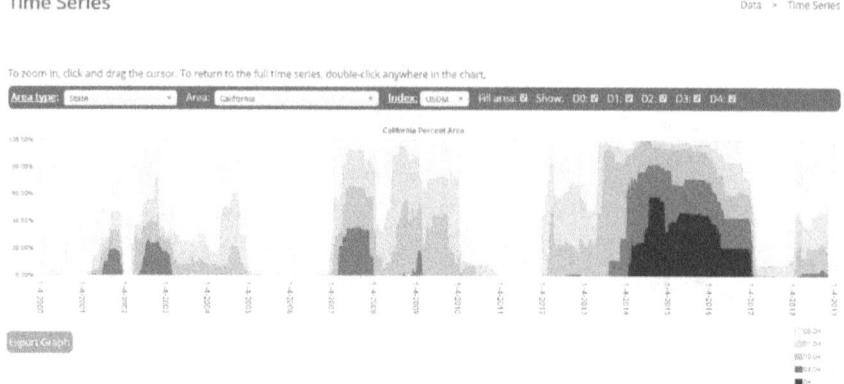

Figure 4.2 — Drought timeline for California for 2000–2018. For a few short years, Californians suffered through a drought that seemed as though it might go on forever, but it didn't.

National Drought Mitigation Center:

Time Series   Data > Time Series

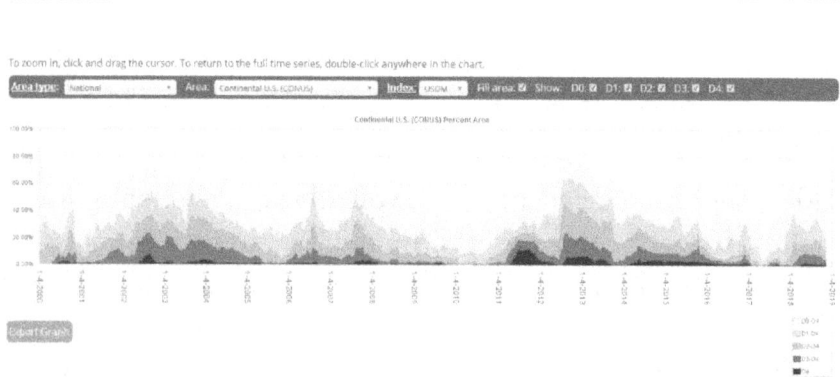

Figure 4.3 — Drought timeline for the Continental United States, 2000–2018.

The corporate media and politicians ignored the idea that climate typically changes in cycles. Alarm sells and garners votes. Like the Ice Age alarm of the 70s, projections were unrealistically linear — forever in one direction. Then global warming returned. Context had been left out. The fact remains that our scary warming is in actual fact, Earth pulling away from the bottom of our planet's livable temperature range—still far below average.

Native American Prosperity

Reid Bryson and Thomas Murray wrote in their 1977 book, *Climates of Hunger,* that archaeological evidence shows great prosperity amongst Native Americans during the Medieval Warm Period. At a site called Mill Creek Site B, the natives had a thriving agricultural economy from about 900 to 1400 AD. This site is centered in the modern corn belt area of northwest Iowa. During that same period, there were as many as 1,000 villages spread across the Great Plains, from Iowa to Colorado. Four hundred years later, when European explorers entered the region for the first time, there were no corn-farming villages; they had all disappeared. Little Ice Age cold had made it too dry to grow crops. All of that earlier prosperity had disappeared because of the colder climate.

Green Sahara

The early part of our Holocene interglacial was far warmer than today. All that extra warmth created enough evaporation from the oceans for significantly more rain. The changes in climate altered the flows of air and ocean currents enough to create a monsoon pattern which drove rain clouds into the Sahara, making it green. This lasted an estimated 3,000 years. Deep into the Sahara, we have signs of human habitation with pictographs of large domesticated herds. Geologists have uncovered evidence that today's wimpy Lake Chad was, back then, a robust inland sea rivaling the Caspian in its heyday. Several other lakes dotted the Saharan landscape from the far more frequent rains.

There is no guarantee that a desert region will develop a similar monsoon pattern, but increased evaporation and rainfall increases the chances that some deserts will disappear or shrink in size.

Conclusion

Land only gets water because of warmth—warming the oceans, lakes, rivers and land—evaporating water. The more warming, the more rain or other precipitation there will be. Within a certain temperature range, warming is a good thing and results in a greener planet, full of life. Right now, we are near the bottom of Earth's livable temperature range. We have plenty of room to warm up before there is cause for concern. All other complaints about warming are similarly flawed. For more on that, see my book, *Climate Basics: Nothing to Fear*.

Change frequently creates problems for some and benefits for others. And while climate always changes, we need to be ready to help one another. But we also need to take all changes in their proper perspective. Warming leads to fewer, shorter droughts and smaller, drier deserts.

Our Modern Warm Period has seen far more prosperity than at any other period in human history. The ingenuity of man, plus greater, life-giving warmth, and more, life-affirming carbon dioxide have us producing enough food to feed 11 billion people. It is only because of pitiful management and the selfishness of a few that those resources are not helping the poorest amongst us.

If we were truly wise, we would see the cycles of the past and the tell-tale clues that help us see the lies being told to us. We would prepare for the coming cold or we would find a way to end the current Ice Age once and for all. Either of these would be the truly compassionate thing to do.

Other books to consider:

More Books on Climate

Climate Basics: Nothing to Fear by Rod Martin, Jr. Debunking the key claims of warming alarmism. #1 Bestseller on Amazon for Science & Math short reads, and #2 Bestseller in Weather.

Thermophobia: Shining a Light on Global Warming by Rod Martin, Jr. In-depth exploration of the "fear of warmth" epidemic.

Red Line — Carbon Dioxide: How humans saved all life on Earth by burning fossil fuels by Rod Martin, Jr. Setting the record straight on the obscene slander against this vital gas of life.

Self-Help and Self-Empowerment

The Spark of Creativity: How to Unleash a Flood of Ideas That Matter, Right Now by Rod Martin, Jr. Overcoming writer's block and every other possible barrier to your own creativity.

Taking Charge: How to assert positive control over your own emotions by Rod Martin, Jr. It's all about discovering the reins of your own life.

Discount available on most ebook titles at
http://TharsisHighlands.WordPress.com/books/.

Appendix

- References
- Notes
- Glossary
- Videos
- Links to Illustrations
- About the Author
- Other Books
- Connect

References

ABC News (Australia). (2018:0803). "Drought stricken farmers are looking for leadership and action on climate change." Retrieved on 2018:1115 from https://youtube.com/watch?v=Ub3RhxUa6Vk

Ault, T., St. George, S. (2018). "Unraveling the mysteries of megadroughts." Retrieved on 2018:1203 from https://physicstoday.scitation.org/doi/10.1063/PT.3.3997

Bastasch, M. (2015:0504). "Flashback 1976: Scientists Blamed California Drought on Global Cooling." Retrieved on 2018:1203 from https://dailycaller.com/2015/05/04/flashback-1976-scientists-blamed-california-drought-on-global-cooling/

Berghuijs, W.R., Aalbers, E.E., *et al.* (2017:1116). "Recent changes in extreme floods across multiple continents." Retrieved on 2018:1208 from http://iopscience.iop.org/article/10.1088/1748-9326/aa8847

Brooks, C.E.P. (1949). *Climate Through the Ages: A Study of the Climatic Factors and Their Variations*. McGraw-Hill Book Company, Inc., New York.

Bryson, R. and Murray, T. (1977). *Climates of Hunger.*
  University of Wisconsin Press, Madison.
CBC News. (2018:1010). "Is Canada doing enough to avoid a
  'climate catastrophe'? | Power & Politics." Retrieved on
  2018:1115 from
  https://youtube.com/watch?v=T9kLGwKbsBI
CNN. (2018:1007). "New climate change report issues stark
  warning." Retrieved on 2018:1115 from
  https://youtube.com/watch?v=RvaCM1TNBBk
Curry, A. (2014). "NBC Documentary. 2014 The year of
  extremes." Retrieved on 2018:1115 from
  https://youtube.com/watch?v=CRZ0KIH1Nls
Deaton, J. (2018:0411). "California's snow drought is a recipe
  for danger." Retrieved on 2018:1205 from
  https://popsci.com/california-snow-drought-climate-
  change
ebay.com. (ND). "EQUICALASTROBUS PINE CONE FOSSIL,
  SAHARA DESERT, MOROCCO, EOCENE." Retrieved
  on 2018:1124 from
  https://ebay.com/itm/EQUICALASTROBUS-PINE-
  CONE-FOSSIL-SAHARA-DESERT-MOROCCO-
  EOCENE-/113384500916
Heim, Richard R. (2018:0103). "A Comparison of the Early
  Twenty-First Century Drought in the United States to
  the 1930s and 1950s Drought Episodes." Retrieved on
  2018:1124 from
  https://journals.ametsoc.org/doi/abs/10.1175/BAMS-D-
  16-0080.1
Heller, T. (2018:0711). "SCIENCE: Symptoms of Global
  Cooling and Global Warming Are Identical." Retrieved
  on 2018:1203 from
  https://realclimatescience.com/2018/07/1981-

climatologists-now-blame-recurring-droughts-and-floods-on-a-global-cooling-trend/

Henderson, G. (2014). "The Dilemmaof Reticence:Helmut Landsberg, Stephen Schneider, and public communication of climate risk, 1971-1976." Retrieved on 2018:1214 from http://meteohistory.org/wp-content/uploads/2015/01/HOM6_06_Henderson_Dilemma_of_Reticence.pdf

Hurlbut, C.S., ed. (1976). *The Planet We Live On: An Illustrated Encyclopedia of the Earth Sciences*, Henry N. Abrams, Inc., New York

Johnston, Ian. (2017:0327). "Climate change is altering global air currents – increasing droughts, heatwaves and floods." Retrieved 2018:1128 from https://independent.co.uk/environment/climate-change-global-air-currents-drought-heatwaves-floods-global-warming-michael-mann-arctic-a7651581.html

Landsberg, H. (1962). *Physical Climatology*. Gray Printing Co., Inc., DuBois, Pennsylvania

Mann, M. (2018:0806). "Climate change is making wildfires more extreme. Here's how." Retrieved on 2018:1128 from https://pbs.org/newshour/show/climate-change-is-making-wildfires-more-extreme-heres-how

Mann, M., Gleick, J. (2015:0323). "Climate change and California drought in the 21st century." Retrieved on 2018:1128 from http://pnas.org/content/112/13/3858

Morano, M. (2015:0914). "New studies & data reveal 'global warming' NOT behind California drought – Not 'unprecedented' – 1970's droughts blamed on 'global cooling'." Retrieved on 2018:1203 from http://climatedepot.com/2015/09/14/new-studies-and-historical-data-reveal-global-warming-is-not-behind-

california-drought-1970s-droughts-blamed-on-global-
cooling/

Mudelsee, M., et al, (2003:0911). "No upward trends in the
occurrence of extreme floods in central Europe."
Retrieved on 2016:0618 from
http://nature.com/nature/journal/v425/n6954/full/nature
01928.html

Mudelsee, M., et al. (2004:1202). "Extreme floods in central
Europe over the past 500 years: Role of cyclone
pathway 'Zugstrasse Vb'." Retrieved on 2016:0618 from
http://onlinelibrary.wiley.com/doi/10.1029/2004JD00503
4/full

National Drought Mitigation Center. (ND). "United States
Drought Monitor, Time Series." Retrieved on 2018:1118
from
https://droughtmonitor.unl.edu/Data/Timeseries.aspx

PBS Newshour. (2018:1008). "World needs to make near-
revolutionary change to avoid imminent climate
disaster. Is there hope?." Retrieved on 2018:1115 from
https://youtube.com/watch?v=uUeOApeSuHU

Random House. (1983). *The Random House Encyclopedia.*
Random House, New York.

ReasonTV. (2009:0310). "MIT Climatologist Richard Lindzen
on the Politics of Global Warming." Retrieved on
2018:1203 from https://youtube.com/watch?v=OS-
cLp1PEGQ

Science Daily. (2018:0423). "Climate change intensifies
droughts in Europe." Retrieved on 2018:1124 from
https://sciencedaily.com/releases/2018/04/180423110822.
htm

Union of Concerned Scientists. (ND). "Causes of Drought:
What's the Climate Connection?" Retrieved 2018:1124
from https://ucsusa.org/global-warming/science-and-

impacts/impacts/causes-of-drought-climate-change-
    connection.html
United States Drought Monitor. (ND). "United States Drought
    Monitor." Retrieved on 2018:1115 from
    https://droughtmonitor.unl.edu/CurrentMap.aspx
University of California Museum of Paleontology. (ND). "The
    Eocene Epoch." Retrieved on 2018:1124 from
    http://ucmp.berkeley.edu/tertiary/eocene.php
Warnert, Jeannette. (2014:0327). "The California drought is
    helping return the weather pattern to normal."
    Retrieved on 2018:1204 from
    https://ucanr.edu/blogs/blogcore/postdetail.cfm?postnu
    m=13274
Wikipedia. (ND). "Eocene." Retrieved on 2018:1124 from
    https://en.wikipedia.org/wiki/Eocene

Notes

Introduction: Some Like it Hot

(No notes for the introduction.)

Chapter 1: Bogus Claims About Deserts and Droughts

Confirmation Bias—Confirmation Everywhere

In science, we need our terms and their definitions to be precise. We rely more on numbers and mathematics than on adjectives to describe what we observe. We need to be able to replicate the claims, but those claims also need to be falsifiable. This merely means that we need to be able to test our ideas.

Scientists and the big corporate news use the term "climate change" as a source of just about everything—warming, cooling, droughts, floods and every other kind of change. This means that just about everything observed confirms its existence. Confirmation bias is built-in. Because of this, "climate change" as a source of observable effects, is not falsifiable. Therefore, it is anti-science.

This sloppy use of terms isn't restricted to news reporters. Scientists have been corrupted with the same sloppy terminology and muddy thinking. Gavin Schmidt of NASA /

GISS had this to say for PBS Newshour, "Basically, this report [IPCC 2018, 4th quarter] is telling us things that scientists have known for a long time—that climate change is already occurring. And it really doesn't take very much more for it to become a very, very serious issue...." Incredibly, this high-profile scientist makes it sound like climate change is something new. It isn't. Far from it; climate has changed for nearly 4.5 billion years—ever since Earth gained an atmosphere. To talk about climate change as if it were something incredibly dangerous is similar to talking about anything else that is ordinary as if it were a grave threat to life on this planet. "Air is already happening." "Time is already happening." "Life is already happening." Are you scared, yet?

When a politician, like former president Obama, says that "climate change is real," they reveal their ignorance and lack of critical thinking skills. "Climate change" is exactly the wrong term to be using about the supposedly scientific effect they fear.

Scientific Consensus: Oxymoron that Stops Critical Thinking

Albert Einstein despised peer review and so-called "scientific consensus" For him, a hypothesis and its supporting evidence should be seen by the scientific community at large. That's the real peer review. The current practice of hiding reviewers, giving them the power as gatekeepers, makes science highly corruptible. In the 1930s, one hundred German scientists published a book condemning Einstein and his Relativity. Einstein commented that all it takes is one scientist with the right facts to disprove a thesis. Group consensus is not needed.

But the problem with consensus thinking runs even deeper. Science run by popularity (politics or voting) ceases to be science. Consensus only gets in the way of the restraint and humility that promote discovery. Some fans of consensus reply with witty quips like, "If I have cancer, I'd prefer to go to a real doctor, than to ask a plumber what I should do." The problem with this logically fallacious thinking is that groups of experts have been wrong before. Entire fields have been wrong and ego made it difficult for science to wake up to that fact.

If doctors still thought bloodletting, mercury and cocaine were viable treatments, many more patients would be dying today. As it is today, doctors use highly toxic and carcinogenic substances to kill the cancer cells, but all too often they kill the patient, too. Big corporations are greedy. They have to be. They have a fiduciary obligation to be selfish in the extreme—enough so to lobby the government to have their competition outlawed. Mention the word "cure" and you could be fined or jailed. Disease maintenance is far more profitable than cures. The mindset of the bloodletting days has never left us. Add to that the powerful motivation of corporate greed, and consensus becomes a self-perpetuating beast.

For decades, American anthropologists were shackled with another point of "scientific consensus" sometimes referred to as the "Clovis First" dogma. The leaders of the field made it difficult for others to change this worshiped consensus—using ridicule, threats of lost funding, and even threats of lost careers. It should be obvious to any thinking person that ridiculing a scientist who dares dig below the Clovis horizon (the strata of earth dated to early Clovis) would tend to suppress any discovery that would jeopardize that dogma. This is not science. It sounds more like a religious cult. This is science forgetting to be skeptical of their own skepticism.

Any and all knowledge is up for review and possible replacement. To think otherwise is to claim an argument from ignorance type logical fallacy. This happened in the late 19th century, when scientists found that Newton's Laws of Motion were imperfect at speeds approaching the velocity of light. It took Einstein's Relativity to resolve that enigma.

This problem of consensus thinking is so big that scientists have been known to reject evidence in favor of tradition. Read that again. A scientist rejected evidence to protect the consensus. That is "slap-in-the-face" stupid. It's certainly not science. And it reveals a huge flaw in science— that of ego and the tendency of some scientists to use muddy thinking when things are too new or unsettling to their own worldview.

Settled Science—The Mindless Attitude of 'Shut Up!'

When one group of people have an agenda they want to protect, ridiculing their opposition is a standard weapon, but it's not logical or ethical. Telling others that all investigation has already been done is like telling others to "shut up and get with the program." Like "scientific consensus," it stifles discussion.

But can science ever be called "settled?" Quite simply, "No!"

First of all, we don't know what we don't know. We don't know what future discoveries may hold. To claim any field of inquiry to be settled is an appeal to the stone logical fallacy. It's also an implied argument from ignorance logical fallacy. To claim that we know everything on a topic is arrogance (a form of cognitive blindness).

On the topic of climate, there are many things we still don't know. Though we understand a great deal about droughts and the changing sizes of deserts, we still don't know everything. We still cannot predict with high confidence when a drought will occur at any one location. And since polar glaciers are one form of desert, they are one aspect of our discussion of droughts. We still don't know what all controls the onset of glacial periods in the current Ice Age.

The Very Unscientific IPCC

The Intergovernmental Panel on Climate Change (IPCC) is a part of the United Nations. The name itself reveals its political nature. Science and politics almost never mix well, especially when there is an agenda to serve.

The IPCC is a thin veneer of science covered over with lots of political stickers. The science is lost in the noise of government motives.

But the media portrays the IPCC differently. One PBS reporter on their News Hour said, "As we reported earlier, the Intergovernmental Panel on Climate Change, which is a consortium of climate scientists, announced today that if the world community doesn't reduce carbon emissions, drastically, millions of people across the planet will suffer dire consequences." All too frequently, what the scientists determine is rewritten by the political hacks and activists who also work for the IPCC.

Modern Warming Faster and More Extreme?

Modern warming is not faster or more extreme than climate changes of the past.

Rates of warming involve degrees of temperature per some unit of time. For this short discussion, let's use degrees Celsius per century. The most quoted rate feared by the UN's IPCC is 3°C per century. This doesn't seem like much, but we're talking about the entire planet.

There are two issues, here. One is the stress on individual organisms; the other involves changes in weather patterns.

Complicating the issue is the uneven distribution of temperature changes across the planet. A 3°C rise in global temperature will have something like a half degree change, or less, in the tropics. That's virtually nothing, especially for a century. The desert subtropics will experience perhaps three-quarters of a degree of warming. Still not much. Mid-latitudes might get a degree or two of warming. That will give places like Minneapolis and New York slightly warmer summers and less harsh winters. Most of the warming will end up in the polar regions, raising temperatures there by as much as 10°C. That will make some of the polar region far more livable. More life will invade those largely dead lands.

The stress on individual organisms is laughably simple. Talking about temperature alone, organisms are far more resilient than the Warming Alarmists make them out to be. Even if a species lives for a full century, each animal or plant won't consciously detect the century of warming. The daily warming in places like Los Angeles is far greater. The largest city in California enjoys nearly 11°C of warming every day— sometimes more. Because there are 36,525.6 days per century, the rate per century turns out to be a whopping 398,129.04°C per century. That's how steep the warming rate is each day in the city of "Angels." Plants and animals are not keeling over from this gargantuan rate of warming. In Phoenix, Arizona, the daily rate is far steeper, and people keep moving there.

The 20th century rate of warming was 0.9°C for the entire century. But that's nothing compared to the warming experienced in the two-year recovery period after the 1816 year without a summer. That would equate to 50°C per century. And, according to some scientists, the warming at the end of the Younger Dryas (about 9620 BC) was equivalent to 100°C per century for that first decade.

So, the rate of warming is not the problem. But neither is the amount of warming. Our modern warming for the last century and a half is relatively weak and slow.

Climate has always changed and so have the weather patterns driven by climate. Because we live near the bottom of

Earth's livable temperature range, warming is far from a problem. Cooling is far more dangerous, and that's just what Bill Gates and Harvard University activists are planning to do—cool the planet by blocking the sun. Former CIA Director Brennan praised techniques like Stratospheric Aerosol Injection to cool the planet "like volcanoes do," not realizing that his talk was on the 200th anniversary of just such a volcanic cooling that killed thousands and turned thousands more into climate refugees (ref. CFR, June 2016).

The Overpopulation Scare

Since the late 1700s, doomsayers have been trying to scare others with predictions of civilization's collapse with the weight of too many humans. About the time Thomas Malthus first published (anonymously) his infamous treatise, "An Essay on the Principle of Population," the Earth was on the cusp of its first billion inhabitants. Malthus warned that over-population would double every 25 years, but that food production would grow only in a linear fashion. The result, he predicted, would be massive starvation unless some kind of birth control measures were implemented. Obviously, he was wrong. It took 127 years to double to 2 billion, 47 years to double again to 4 billion, and is estimate to double again in 49 years to 8 billion by 2023. Yet, even with all this doubling, food production has kept pace with the need. In fact, food production has been estimated by some to be enough currently to feed 11 billion. Greed and mismanagement make the proper distribution of this food somewhat problematic. Too many prosperous people are now so dependent upon governments that they've lost all sense of individual compassion. And governments are not doing what's needed to take up the philanthropic slack.

Exactly 170 years after Malthus's population book was first published, Paul Ehrlich published his *Population Bomb,* with similarly grave warnings about the future—predictions which similarly flopped, big time. These social engineers and collectivists like to think they know better for the rest of us humans. They want to run our lives—for our own good. But we don't want their micromanaging, back seat driving in our lives. Human ingenuity has outstripped their expectations, making food production more and more efficient.

One enterprising researcher calculated that all of the humans on planet Earth (7.0 billion at the time) could be squeezed, shoulder-to-shoulder into the space of New York's five boroughs. That would leave the rest of New York state entirely empty of people. In fact, it would leave virtually all of the planet empty of humans.

A more realistic scenario had each human family—an average of 4 people—in modest, two-story homes crowded into the state of Texas. Again, the rest of the world would be empty of people.

With each family growing their own food in a roof top garden, with aquaculture fish tank under the lettuce and tomatoes, Earth could sustain over 400 billion humans with plenty of room for nature. Such a landscape would be packed, for sure, but it would all be based on today's technology, Earth's current resources and a sane, unselfish and tiny government. With futuristic technology, the stable population limit may be far higher—possibly into the trillions, like the galaxy's capital planet in *Star Wars*.

Today's globalists want power and control over our lives. They don't like life—messy and unpredictable. Again, overpopulation has never been a problem for humanity.

Chapter 2: Counter-Claim — How Rain Benefits from More Warming

(No notes for this chapter.)

Chapter 3: How it All Works — Rain, Droughts and Deserts

(No notes for this chapter.)

Chapter 4: Empirical Evidence About Deserts and Droughts

(No notes for this chapter.)

Glossary

Note: Not every term or concept has been included in this glossary. I encourage you to explore the subject online or in books on those terms for which you would like more information. Make learning a lifetime occupation.

carbon dioxide *n.* — an odorless, colorless gas and a minor constituent of the Earth's atmosphere. Without this trace gas, all life on Earth would die. Frequently abbreviated CO_2. This is what plants breathe. And plants "exhale" oxygen (which see). Not to be confused with poisonous carbon monoxide (CO).

climate *n.* — a persistent average state of a region's weather, typically taken over a period of several decades — usually thirty years.

climate change *n.* — modification of a region's persisting average weather. This can include warming or cooling, alterations in turbulence, patterns of flow, timing of events, atmospheric chemistry and more. Such modifications have occurred throughout the existence of our planet's atmosphere — more than 4 billion years. This term has been kidnapped by a modern movement to

mean only "recent, manmade, warming and catastrophic" modifications to the atmosphere. This is a
distortion of the original definition.

CO_2 *n.* —carbon dioxide (which see).

drought *n.* —a period of decreased rainfall that is insufficient
for the life forms within a region. Drought typically
occurs from changes in weather patterns, but more
importantly from regional or global cooling. In fact, the
global cooling of the last 50 million years or so has
significantly desiccated the planet, increasing the extent
of subtropical deserts and creating polar deserts.

Earth *n.* —our home planet. It possesses a breathable
atmosphere, water in three key phases (solid, liquid,
gas), dry land and a surface teeming with life. It also
has one natural satellite typically called the Moon.

glacial *n.* —a cooler period of increased glaciation during an
Ice Age in which polar glaciers expand to cover large
portions of adjacent continents. During the last 1.1
million years of the current Ice Age, glacial periods
have averaged 90,000 years in length (ref: W.S. Broecker,
1998). The duration of glacial periods for the last
800,000 years has varied between 24,000 and 143,000
years. Compare *interglacial*.

global cooling *n.* —a decrease in the average temperature of
the planet. This can be a bad thing during our current
Ice Age. Cooling tends to produce less evaporation, and
thus drier climate.

global warming *n.* —an increase in the average temperature of
the planet. This can be a good thing during our current
Ice Age. Warming tends to produce more evaporation,
and thus moister climate.

Holocene *n.* —an interglacial of the current Ice Age; the
current interglacial (which see).

Holocene Optimum *n.*—a period of about 3,000 years wherein the northern hemisphere was as much as 1.1°C warmer than today. This warmth, compared with the shallow cool periods (roughly as warm as today) resulted in a green Sahara for about 3,000 years.

hurricane *n.*—A dangerous tropical cyclone of the Atlantic Ocean region. Compare *typhoon*.

Ice Age *n.*—a period of increased cooling where both polar regions experience permanent glaciation throughout the year. The current such period has had glaciation in Greenland and Antarctica for roughly 2.6 million years. Such periods include several glacial and interglacial periods, alternating between warmer and cooler phases.

interglacial *n.*—a warmer period of relaxed glaciation during an Ice Age in which polar glaciation recedes and global climate warms noticeably. The amount of warming and glacial receding can vary a great deal with some such periods being as much as 5°C warmer than our current Modern Warm Period, or 2°C cooler. There have been several dozen interglacials in the current Ice Age. For the last 1.1 million years, interglacials have averaged about 11,000 years in length (ref: W.S. Broecker, 1998). The duration of interglacial periods for the last 800,000 years has varied between 4,000 and 24,000 years. Compare *glacial*.

Intergovernmental Panel on Climate Change *n.*—a political organization associated with the United Nations tasked with determining the nature and extent of man's impact on the planetary climate as a result of burning fossil fuels.

IPCC *n.*—Intergovernmental Panel on Climate Change (which see).

Jupiter *n.*—the largest planet in our star system, roughly ten times the diameter of Earth, with a thick atmosphere many thousands of kilometers deep and no solid surface. Because of its great distance from the sun, it is extremely cold at the tops of the clouds, near the air pressure equivalent to that on Earth's surface. Despite the extreme cold, the planet hosts some of the largest storms in the solar system.

oxygen *n.*—a key constituent of Earth's atmosphere and the most vital gas for animal life. Animals exhale carbon dioxide (which see).

parts per million *n.*—the concentration of something as a fractional measure compared to a whole. If you take a million of something, the count given will be the number of pieces or portions out of that whole million that apply to a specific substance. This is similar to the term percent. Example: The atmosphere consists of 21 percent oxygen, or 210,000 parts per million oxygen.

Pleistocene *n.*—the current Ice Age (which see). This period of permanent polar glaciation has lasted for 2.6 million years. Before scientists knew very much about Earth's history, they thought the Pleistocene ended 11,600 years ago. Today, we know that the current epoch—the Holocene—is merely one in a series of dozens of interglacial periods that are part of this Ice Age.

ppm *abbr.*—parts per million (which see).

tornado *n.*—a small, cyclonic storm, typically less than several hundred meters across, with extremely fast winds and an ability to create tremendous damage to buildings and to anything else above ground.

typhoon *n.*—a dangerous tropical cyclone of the Pacific Ocean region. Compare *hurricane*.

Venus *n.* — our sister planet, closer to the sun. The planet is slightly smaller than Earth (which see), has a crushing atmosphere of mostly CO_2, a heavily reflective cloud cover, and a surface with virtually no wind and temperatures hot enough to melt lead (462°C). The planet spins very, very slowly and has no natural satellite.

warm period *n.* — a span of time which has a higher temperature than the preceding and succeeding spans of time. Climate always changes and most frequently in repeating cycles. The Holocene has contained 10 clearly-defined major warm periods on a roughly 1,000-year cycle. Cycles of other periods make the pattern of warming and cooling more complex than they would be if there were only one cycle involved. The most recent four major warm periods of the Holocene have been, the Modern (1850 to today), the Medieval (850–1350), the Roman (200 BC–AD 100) and the Minoan (1400–1100 BC).

Link to Illustrations

For those who would like to see the pictures in color or larger in size, the illustrations have also been made available online at,
https://tharsishighlands.wordpress.com/2018/12/14/illustrations-used-in-deserts-droughts/

Videography

Be sure to Like, Comment and Subscribe to the channel: https://youtube.com/c/RodMartinJr/

Climate Change Lies Exposed series

Top 10 Climate Change Lies Exposed (*over 250k views*)
https://youtube.com/watch?v=ICGal_8qI8c
Climate Change Lie #1 Exposed: Global Warming is Bad
https://youtube.com/watch?v=KbfjEPo083U
Climate Change Lie #2 Exposed: CO2 Causes Dangerous Global Warming
https://youtube.com/watch?v=ZH5ATcpMJQo
Climate Change Lie #3 Exposed: Global Warming Causes Extreme Weather
https://youtube.com/watch?v=aTiBbAGl0qI
Climate Change Lie #4 Exposed: Global Warming causes droughts
https://youtube.com/watch?v=DusZ5dP4hDw
Climate Change Lie #5 Exposed: Our current warmth is unusual
https://youtube.com/watch?v=FR2aZc5bjUU

Climate Change Lie #6 Exposed: Our current level of CO2 is unusually high
https://youtube.com/watch?v=ASV3UUwYZg0
Climate Change Lie #7 Exposed: The rate of warming is dangerous
https://youtube.com/watch?v=OsJ67Hp4l-g
Climate Change Lie #8 Exposed: The Science is Settled
https://youtube.com/watch?v=6yzkAjWY8rM
Climate Change Lie #9 Exposed — There is a consensus on dangerous, man made, Global Warming
https://youtube.com/watch?v=URE4NMk1DbA

Carbon Dioxide Fan Club

Earth vs. Venus: Will our world ever suffer runaway greenhouse warming?
https://youtube.com/watch?v=SO1M8GEDyYk
Top 10 Facts that Prove CO2 Does NOT Drive Global Temperature
https://youtube.com/watch?v=CSQlJx76b64
Verdict: CO2 Not Guilty! Greenhouse DESTROYED! Must see!
https://youtube.com/watch?v=1f6zB320Hac

Global Warming Fan Club

How Global Warming Made Civilization Possible
https://youtube.com/watch?v=057GgxpZWRc
Top 10 Reasons Global Warming is Good
https://youtube.com/watch?v=dQc4iXgrrEo

Big Climate Quiz (BCQ)

Climate Change Smarts #1: Why didn't civilization start during the last glacial period?
https://youtube.com/watch?v=Bf0gty2XAjw
Climate Change Smarts #2: What Causes Wind to Blow?

https://youtube.com/watch?v=lhk7JIQ6e-U
Climate Change Smarts #3: How does land ever get water?
https://youtube.com/watch?v=do0kb7Udq-g
Climate Change Smarts #4: What is an Ice Age?
https://youtube.com/watch?v=RjMbE-G8JFo

Climate Music Video series

Thermophobia - Why Fear of Warming in the current Ice Age is all wrong
https://youtube.com/watch?v=Q68fIkdC9Rk
Extreme Weather - How the Climate Change Alarm is All Wrong
https://youtube.com/watch?v=x18gwLpLI2A
Thermophobia -- Debunking: "Global Warming causes more storms"
https://youtube.com/watch?v=d40_2yGuV_o

About Rod Martin, Jr.

Rod Martin, Jr. is a modern polymath (Renaissance man)—artist, scientist, mathematician, engineer and philosopher. He first became interested in climate science in the mid-70s. A forest ecology PhD friend of his was retiring and donated two climate texts to the cause. Initially, Martin's interest in the subject covered planetary atmospheres—weather systems, atmospheric retention rates, optical thickness (greenhouse effect), adiabatic lapse rates, climate chemistry and planetary habitability.

Like so many others, during the 70s, 80s and 90s, Martin's interest in ecology and the environment continued to grow. When Al Gore's film, *An Inconvenient Truth,* came out in 2006, Martin was an immediate fan. But as the controversy on the topic heated up, Martin suddenly realized that all of the things he had learned about climate over the years contradicted many of the so-called facts in Gore's award-winning film.

In college, from the mid-90s to the early 2000s, Martin studied computer science, earning a degree, *summa cum laude.*

With only a 139 IQ, Martin realized that he was not the sharpest implement in the tool shed. In fact, all of his younger brothers had far higher IQs. From this relative handicap, he learned the immense value of humility and the need to remain unattached to any ideas, lest they become dogma, and blind him from further discovery. Thus, he was able to learn the true value of skepticism, and was able to recognize the inevitable pitfalls of that scientific paradigm. He also made the distinction between confidence in knowledge (an enormous source of blindness) and confidence in one's ability to find new knowledge (a source of empowerment).

In 2016, Martin implemented a campaign to set the record straight on climate. He wasn't alone. Many climate scientists, astrophysicists, meteorologists and concerned citizens had already begun to speak out against the so-called "scientific consensus" (an oxymoron, because science is never done by consensus). Martin has created numerous educational videos on climate change and global warming, and created a website to discuss these topics in greater detail.

https://GlobalWarmthBlog.WordPress.com/

From a lasting love of stars and astronomy, he created 3D space software, "Stars in the NeighborHood," available online.

https://SpaceSoftware.WordPress.com/buy-now/

He currently resides in the Philippines with his wife, Juvy.

He has taught mathematics, information technology, critical thinking and professional ethics at Benedicto College, Mandaue City, Cebu. He continues to teach online and to write.

Other Books by Rod Martin, Jr.

Non-Fiction (as Rod Martin, Jr.)

The Art of Forgiveness, Tharsis Highlands (2012, 2015)

The Bible's Hidden Wisdom: God's Reason for Noah's Flood,
Tharsis Highlands (2014)

The Spark of Creativity, Tharsis Highlands (2014)

Dirt Ordinary: Shining a Light on Conspiracies, Tharsis
Highlands (2015)

Favorable Incompetence: Shining a Light on 9/11, Tharsis
Highlands (2015)

Thermophobia: Shining a Light on Global Warming, Tharsis
Highlands (2016)

*Red Line—Carbon Dioxide: How humans saved all life on Earth by
burning fossil fuels*, Tharsis Highlands (2016)

*The Science of Miracles: How Scientific Method Can Be Applied to
Spiritual Phenomena*, Tharsis Highlands (2018)

Proof of God, Tharsis Highlands (2018)

Deserts & Droughts: How does Land Ever Get Water?, Tharsis
Highlands (2018)

*Taking Charge: How to Assert Positive Control Over Your Own
Emotions*, Tharsis Highlands (2018)

Spirit is Digital — Science is Analog: Discovering where miracles and logic intersect, Tharsis Highlands (2019)

Proof of Atlantis? Evidence of Plato's Lost Island Empire, Tharsis Highlands (2019)

Enemies of Christ: The Need to Protect Our Own Salvation from Ravening Wolves, Tharsis Highlands (2019)

Science Fiction (as Carl Martin)

Touch the Stars: Emergence, with John Dalmas, Tor (1983), *expanded* Tharsis Highlands (2012)

Touch the Stars: Diaspora, Book 2 of Touch the Stars, Tharsis Highlands (2014)

Entropy's Children, anthology of short fiction, Tharsis Highlands (2014)

Gods and Dragons, Book 1 of *Edge of Remembrance,* Tharsis Highlands (2017)

Tales of Atlantis Lost, Book 2 of *Edge of Remembrance,* Tharsis Highlands (2017)

An Excerpt from *Thermophobia*

Introduction: Thermophobia in Perspective

"One of the brightest gems in the New England weather is the dazzling uncertainty of it." — Mark Twain, speech to the New England Society, December 22, 1876

Thermophobia is a fear of warmth — a disease which has gripped much of the planet in recent years. Every fear about global warming will be cured in this book. We will use facts, logic, science and a sense of perspective that has been sorely missing from much of the mainstream media. Don't misunderstand this. Caring about the planet and our future is the foundation of this book.

Let us start with some simple questions.

Do you fear cuddling? Do you shake in terror over the idea of wrapping up in a warm blanket on a cold, rainy or snowy night? Does the notion of a warm bath or shower send chills up your spine? And do you stand frozen in horror thinking about sitting down to a nice, hot meal?

No?

Then why has civilization gone suddenly bonkers over a rise in average global temperature of 5.4 °F (3 °C) *over the next century?* That's a temperature difference some people would be hard pressed to discern. That's a yearly increase of only 0.06 °F (0.035 °C), barely enough to be measurable by most and felt by none. When a similar, natural rate of warming delivered us from the frozen horrors of the Little Ice Age, no one complained.

Let's do some comparison.

Downtown Los Angeles has a coastal, Mediterranean-type climate with an average daily high in the rainy month of February of 68.6 °F (20.3 °C) and an average daily low of 49.3 °F (9.6 °C). This gives us a daily range of about 19.3 °F (10.7 °C). *Frightening!* That's more than three times the dreaded increase foretold by the United Nations. But the people in Southern California don't seem to mind. And that's once a day, every day in February. And that's a minor fluctuation, dampened by the weather and proximity to the Pacific Ocean.

Phoenix, Arizona has an inland, high desert climate with an average daily high in dry June of 103.9 °F (39.9 °C) and an average daily low of 77.7 °F (25.4 °C). This gives us a daily range of a whopping 26.2 °F (14.5 °C). *Terrifying!* And yet Phoenix has been such a magnet, its population has grown from nearly 107,000 in 1950 to over 1.4 million in 2010—an increase of nearly 14 times. When my younger brother, Terry, worked for the US Postal Service, he told me they were opening up a new route or two every month just to keep up.

Over a twelve-month period, every year, temperatures in Los Angeles vary from an average low of 47.5 °F (8.6 °C) to an average high of 84.4 °F (29.1 °C). That's a *mind-numbing* yearly change of 36.9 °F (20.5 °C)—nearly seven times the UN's horror story.

For the yearly changes in Phoenix, we might just *blow a gasket*. Temperatures there vary from an average low of 44.8 °F (7.1 °C) to an average high of 106.1 °F (41.2 °C). That's a yearly change of 61.3 °F (34.1 °C)—every year, over eleven times the "global warming" scare story. People in Phoenix are not freaking out over this amount of yearly temperature change. Maybe they know something the UN doesn't.

Climate is a complex, non-linear system which is globally stable and locally unpredictable. This means that the entire planet will experience changes gradually. Simple inputs, however do not necessarily have simple outputs. Increase the water vapor worldwide, and you will tend to increase the potential rainfall overall, but the rainfall at any one location will remain unpredictable. But even that effect is not simple. Global warming will cause more evaporation from the oceans, but it will also result in the air being able to hold more water vapor before it precipitates as rain or snow. Non-linear simply means that the effects are displayed as curves on graphs rather than straight lines. Some of those curves can be quite complex.

Also, there are many things we do not yet know about the processes and effects within the climate and weather systems. We don't yet know the sensitivity of temperature change with changes in carbon dioxide. The paleoclimatic record tends to show that temperature has a relatively low sensitivity to changes in CO_2, but that's only a qualitative assessment. We don't yet know the exact quantity of this sensitivity. Thus, UN climate models will continue to yield outputs which do not match reality. Also, we do not yet know all of the factors which change climate. For instance, we're still studying the processes by which clouds are formed and their effect on overall climate.

More questions!

Where do most people want to vacation—snow or beach—cold climate or warm? Where do New Yorkers like to retire, if they have the money—Greenland or Florida? Which provides a more hospitable environment for growing crops— cold or warm? Where are we more likely to grow the food to feed millions—Antarctica or Java?

If you've been paying attention to the world around you, likely you will give the warmer answer for each of these questions. This seems to be almost a no-brainer, yet the news and government reports are awash with the horrors of global warming! Thus the ironic term, *thermophobia*—a fear of warmth.

In 2013, according to Mike Schneider of *Business Insider*, an estimated 537,000 residents moved to Florida and 10% of them were from New York state—or about 54,000. An esti- mated zero residents moved to Greenland. Certainly, real estate is far cheaper in Greenland, but facilities and infrastructure are largely nonexistent there. The infrastructure is missing, because the demand is near nonexistent. Warmer climates trump colder ones by a vast margin.

When we compare the population density of various latitudes, we see a clear correlation. Southern United States, including California, Oklahoma, Tennessee, North Carolina and states farther south, have an average density of 49.02 people per square kilometer. Northern states have an average population density of 28.48 people per square kilometer. Southern Canada has 6.82 people per square kilometer, while

Northern Canada drops to 0.16 people per square kilometer. The gradation is clear. People like it warmer.

Below is a map showing the population gradient for the entire world by 10° latitude bands. This is based not on total population, but population per square kilometer of land for each band. Part of the reason for the bias toward the northern hemisphere has to do with the warming effect of the Gulf Stream. That flow of warmer water makes higher latitudes more livable in Europe than they would otherwise be.

In Canada, for instance, the southern border of Northwest Territories stands at 60°N. The closest large city to this latitude is Yellowknife with a population of about 19,000 people. Near that same latitude in Sweden is the capital city, Stockholm. Its metropolitan area has a population of over 2.1 million people. Oslo, Norway has about the same latitude and a metro area population of over 1.7 million.

The amount of warming being predicted as "catastrophic," most people would barely notice. The charts being bandied about by politicians and computer climate modelers show mountains of temperature increase that really are only fractions of a single degree. We're looking at mountains that are visible only under a microscope. The error bars of data sometimes are nearly as large as the swings in the data. In other words, someone is making a big deal out of practically nothing.

https://tharsishighlands.wordpress.com/books/thermop hobia-global-warming/

Connect with Rod Martin, Jr.

Rod Martin, Jr. is his pen name for non-fiction. Carl Martin is his pen name for fiction.
BitChute—https://bitchute.com/channel/M63WrjRpNSPT/
Minds—https://minds.com/RodMartinJr
Gab—https://gab.ai/RodMartinJr
Website and Blog—https://rodmartinjr.wordpress.com/
HubPages—https://hubpages.com/@lone77star
Smashwords author page—
        https://smashwords.com/profile/view/CarlMartin77
Smashwords author page—
        https://smashwords.com/profile/view/RodMartinJr
Udemy courses page—https://udemy.com/user/rodmartinjr/
Facebook—https://facebook.com/RodMartinJr/
Twitter—https://twitter.com/LoneStar77/
Google+—https://plus.google.com/+RodMartinJr/
YouTube—https://youtube.com/c/RodMartinJr/
Goodreads author page—https://goodreads.com/Carl_Martin
Goodreads author page—
        https://goodreads.com/Rod_Martin_Jr

Amazon author page—https://amazon.com/Carl-
    Martin/e/B008CX8KN6/
Amazon author page—https://amazon.com/Rod-Martin-
    Jr/e/B008CZ9JTS/

www.ingramcontent.com/pod-product-compliance
Lightning Source LLC
Chambersburg PA
CBHW072200170526
45158CB00004BB/1721